# YOUR KNOWLEDGE HAS VALUE

- We will publish your bachelor's and
  master's thesis, essays and papers

- Your own eBook and book -
  sold worldwide in all relevant shops

- Earn money with each sale

Upload your text at www.GRIN.com
and publish for free

**Bibliographic information published by the German National Library:**

The German National Library lists this publication in the National Bibliography; detailed bibliographic data are available on the Internet at http://dnb.dnb.de .



**Imprint:**

Copyright © 2015 GRIN Verlag, Open Publishing GmbH
Print and binding: Books on Demand GmbH, Norderstedt Germany
ISBN: 9783668229525

**This book at GRIN:**

http://www.grin.com/en/e-book/323077/active-vibration-control-of-plate-structure-using-electromagnetic-transducer

Zhonghui Wu

# Active Vibration Control Of Plate Structure Using Electromagnetic Transducer Based On H_∞ Optimized Positive Position Feedback

GRIN Publishing

# Table of Contents

Active Vibration Control Of Plate Structure Using Electromagnetic Transducer Based On $H_\infty$ Optimized Positive Position Feedback


Zhonghui WU



**Abstract**

In this paper active vibration control (AVC) methodology is presented by the author using self-sensing magnetic transducers for a flexible plate structure. $H_\infty$ Optimized positive position feedback (HOPPF) controller is tested and verified for multi-modes multi-input-multi-output (MIMO) vibration suppression through simulation and experiment implement. Genetic algorithm (GA) searching is applied to obtain the optimal parameters of the controllers according to the minimization criterion solution to the $H_\infty$ norm of the whole closed-loop system.


Keywords

$H_\infty$ Optimized positive position feedback, Multi-Input-Multi-Output, Genetic Algorithm, Active Vibration Control

## 1. Introduction

Lightweight products and materials were employed by many designers to decrease the cross-sectional dimensions of the structures, improve dynamic performance and operating efficiency. The structures is becoming more flexible and susceptible. The harmful effects of unwanted vibration can be seen easily, especially at the moment the structure is operated at or near their natural frequencies or excited by disturbances which is coinciding with their natural frequencies [1]. Modal control is provided by vibration control engineers to suppress the vibration of flexible structures and has become the best choice for many years. Modal analysis and control refer to extract the interested mode signal from the structural response, decompose the dynamic equations of mechanical system into modal coordinates and design the a single degree-of-freedom oscillator in the modal domain [2,3].

Independent modal space control (IMSC) is proposed by Meirovitch, which can design the controller for each single mode and can be implemented independently to avoid the spillover to the residual modes [4-5]. But according to testing, it needs many sensors/ actuators as the number of modes which need to be controlled, it only can control a limited number of modes and not robust to uncertainties such as parameter fluctuation [6].

Moheimani raised resonant control method, which need to choose a high gain controller at the natural frequency of the flexible structure [7-12]. That controller has a decentralized characteristic from a modal control perspective and roll off quickly at the natural frequency of the structure to avoid spillover [13]. But it has limited method to increase damping to the structure and the performances is restricted [14].

Goh and Caughey provided Positive position feedback (PPF) after compared with other methods [15,16], and the stability was certified by Fanson J. L. [17,18]. PPF controller has several significant advantages and has been applied by many researchers that it is a reliable vibration control strategy to suppress the vibration of flexible systems [15-17, 19-33].

PPF had been modified [34-40] and had showed the robust ability [41-44]. In order to achieve better performance, Genetic Algorithms (GA) was chosen by the researcher to find the placement of sensor and actuator [45-46].

MIMO PPF controller has been verified by [47,14] on a cantilevered beam according to experimental implementation using pole placement and $H_\infty$ optimization method. [48] designed MIMO PPF controller on a flexible manipulator and through the GA method to find controller parameters to optimize $H_\infty$ result. [62] designed and compared HMPPF and HMVPF controller, which is followed [47,14] method to control multi-modes of the beam structure. MPVF controller was designed and provided by [64] for the vibration suppression of the beam.

[49-51] designed and verified GA method for tuning MIMO PPF controller. [52, 54] studied MIMO PPF controller based on the block-inverse technique to suppress the vibration of the grid and shell structure. Using PPF theory [55] gave a decentralized MIMO experimental compensation method for switched reluctance machine.

Through pseudo-inverse technique, [53] designed and proposed MIMO PPF controller. [56] presented the nonlinear MIMO PPF control method for both high and low amplitude vibration suppression of the flexible cantilever plate. MIMO PPF and PD combined controller is verified by [57] for the decoupled bending and torsional modes of the plate. In order to suppress vibration of sandwich plate, MIMO PPF and MIMO SISO were compared by [63].

In this paper, multi-modes multi-input-multi-output (MIMO) $H_\infty$ optimization PPF (HOPPF) controller will be studied by the author and applied the first time to suppress

Figure 2.1 A thin plate in transverse vibration

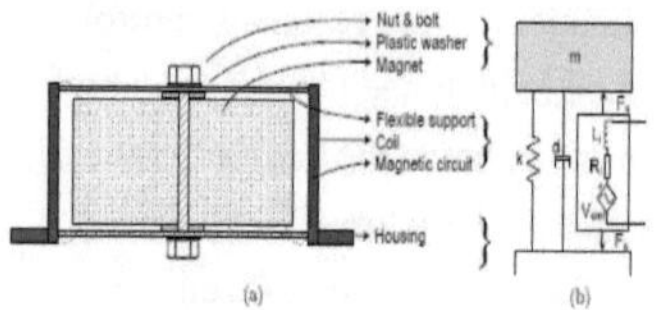


Figure 2.2   Transducer cross section [58]

the vibration of a flexible plate structure.

## 2. Flexible Plate Structure Modeling

### 2.1 Experimental Model

In Fig. 2.1, we can see the schematic of the whole experimental plant t. An MDF board, which is screwed with four rubber legs, is placed on a table. On the MDF board, a uniform AL6061-T6 plate is screwed with three electromagnetic transducers (anticlockwise number 1, 2, 3). Another electromagnetic transducer is mounted as a disturbance noise shaker.

The mechanical the electric circuit model of the transducer are presented in Figure 2.2a, 2.2b separately. The measured mechanical parameters of the transducer are shown in Table 2.1. The transducer consists of a coil, producing a magnetic field when a current is fed through. The core is mounted inside the coil. The magnet of the core will be changed when the vibration is sensed from the plate [58]. The transducer is assumed to be mechanically identical. The natural frequencies of the whole plant can be obtained when the sweep

signal is applied to the disturbance noise transducer.

The amplitude of the vibration is measured by accelerometers and shown on the oscilloscope.

Frequency domain results show that the first four modes, 27.1 Hz, 34.4 Hz, 40.5 Hz, 49.2 Hz, are the dominant modes for all the experimental models.

### 2.2 Analytical Model

This analytical model is including a thin plate and four supporting transducers.

A thin plate is assumed with dimensions: a * b * h. The Young's modulus of elasticity, density and Poisson's ratio of the plate are denoted by E, $\rho$, and v respectively. The deflection is in X and Y directions [59].

The partial differential equation (PDE) of a thin plate under transverse vibration can be written as [59]:

$$\rho(x, y)\, h\, \frac{\partial^2 w}{\partial t^2} + D\nabla^4 w(t, x, y) = \frac{\partial^2 M_x}{\partial x^2} + \frac{\partial^2 M_y}{\partial y^2} + p_z(t, x, y)$$

(2.1)

where

$$\nabla^4 w = \frac{\partial^4 w}{\partial x^4} + 2\frac{\partial^4 w}{\partial x^2 \partial y^2} + \frac{\partial^4 w}{\partial y^4}$$

(2.2)

## 2.3 Modal Analysis

Modal analysis methodology can be applied to find the solution to the PDE (2.1) for uniform thin plates with four supporting transducers under transverse vibration.

The typical PDE for flexible structures can be presented as [1]:

$$\mathcal{L}[w(x, y, t)]$$
$$+ C\left[\frac{\partial w(x,y,t)}{\partial t}\right] + \mathcal{M}\left[\frac{\partial^2 w(x,\ y,t)}{\partial t^2}\right] = f(x, y, t)$$

(2.3)

$\mathcal{L}$ and $\mathcal{M}$ are linear homogeneous differential operators of order 2p and 2q respectively, and $q \leq p$. Here, x, y is the spatial coordinate, which is defined over a domain $\mathcal{D}$. The general arbitrary input is denoted by f, which is distributed over $\mathcal{D}$. The boundary conditions corresponding to PDE (2.3) can be expressed as

$$\mathcal{B}i\,[w(x, y, t))] = 0, \quad i = 1, 2, ..., p$$

(2.4)

where $\mathcal{B}i$ is a linear homogeneous differential operator of order less than or equal to $2p - 1$.

The transfer function of the system can be derived as [1]:

$$G(s, x, y) = \sum_{m=1}^{\infty} \frac{\phi_{mn}(x,y,t)P_m}{s^2 + 2\xi_m \omega_m s + \omega_m^2}$$

(2.5)

The Equation (2.5) is an infinite-dimensional transfer function because there is an infinite number of modes. That is a general solution of PDE (2.3). ($P_m$ is the time-independent forcing term)

The solution, which is for a particular structure, can be solved through finding the eigenfunction ($\phi_{mn}(x, y, t)$), the natural frequency ($\omega_m$), and the structural damping ($\xi_m$) of the structure [1].

The solution to the PDE (2.1) is assumed to be: $w(t, x, y) =$

$$\sum_{m=1}^{\infty} \sum_{n=1}^{\infty} \phi_{mn}(x, y, t)\, q_{mn}(x, y, t),$$

where $q_{mn}(x, y, t)$ and $\phi_{mn}(x, y, t)$ are the generalized coordinate and the eigenfunction, respectively [59].

Based on the orthogonality properties of the eigenfunctions $\phi_{mn}(x, y, t)$, PDE of thin plate with four supporting transducers under transverse vibration can be solved independently for each mode [59].

According to the modal analysis solution, the transfer function from the actuator voltages $V_a(s) = [V_{a1} ... V_{aI}]^T$, to

5

the plate deflection $w(x, y, s)$ can be written as:

$$G(s, x, y) = \sum_{m=1}^{\infty} \sum_{n=1}^{\infty} \frac{\phi_{mn}(x,y,t)\overline{\Psi}_{mn}^{T}}{s^2 + 2\xi_{mn}\omega_{mn}s + \omega_{mn}^2} \quad (2.6)$$

where $(x, y) \in \mathbb{R}$, $\mathbb{R} = \{(x, y) \mid 0 \leq x \leq L_x, 0 \leq y \leq L_y\}$, $\xi_{mn}$ is the damping ratio associated with the mode $(m, n)$.

$\overline{\Psi}_{mn} = [\Psi_{mn1} \ldots \Psi_{mnI}]^{T}$ is affected by the properties of the plate, actuator place and the eigenfunctions $\phi_{mn}(x, y, t)$.

| Parameter | Symbol | Value |
|---|---|---|
| Spring constant | $k$ | 19037 N/m |
| Core mass | $m$ | 0.323 kg |
| Damping ratio | $\zeta$ | $3.39 \cdot 10^{-2}$ |
| Damping Coefficient | $d$ | 5.32 Ns/m |
| Damped resonance frequency | $\omega_d$ | 38.63 Hz |
| Resonance frequency | $\omega_c$ | 38.65 Hz |

Table 2.1: Mechanical parameters

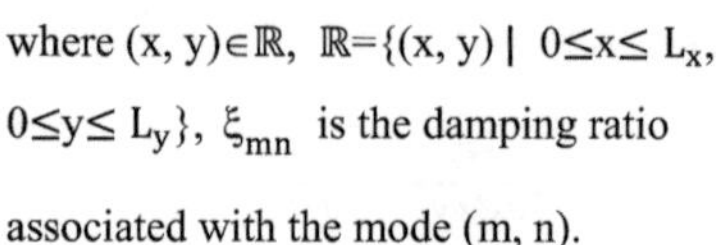

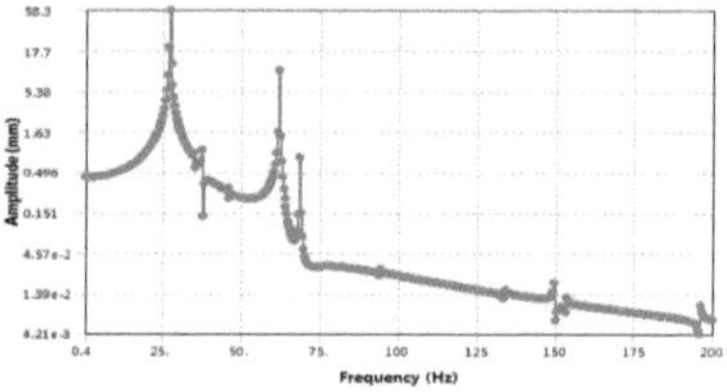

Figure 2.4 Amplitude Response

Figure 2.3   ANSYS Model

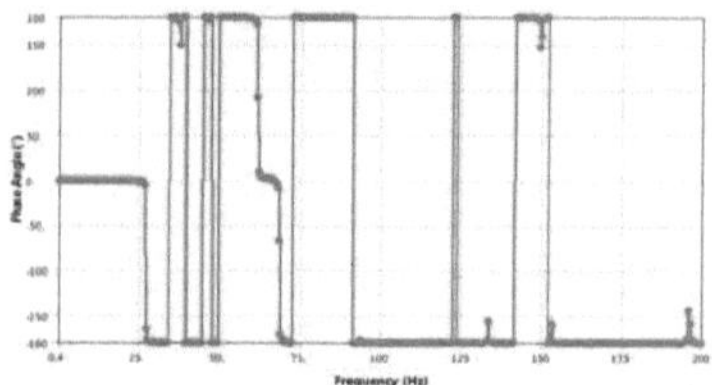

Figure 2.5 Phase Response

## 2.4 Numerical Model

Using software tool ANSYS, numerical model is designed and presented in Fig 2.3

MIMO properties, any output is induced by superposing all the inputs and their related dynamics.

## 2.5 Simulation Model of The Plate

The dynamics of the whole system can be written as output-input open loop system superposition based on the

$$\begin{bmatrix} Y_1(s) \\ Y_2(s) \\ Y_3(s) \end{bmatrix} = \begin{bmatrix} T_{11}(s) & T_{12}(s) & T_{13}(s) \\ T_{21}(s) & T_{22}(s) & T_{23}(s) \\ T_{31}(s) & T_{32}(s) & T_{33}(s) \end{bmatrix} * \begin{bmatrix} U_1(s) \\ U_2(s) \\ U_3(s) \end{bmatrix}$$

(2.7)

where

$$\begin{bmatrix} U_1(s) \\ U_2(s) \\ U_3(s) \end{bmatrix} = \begin{bmatrix} D_1(s) \\ D_2(s) \\ D_3(s) \end{bmatrix} * U_d(s)$$

(2.8)

$U_i$ - input through the transducer i to the top plate, i = 1, 2, 3; $U_d$ - input to shaker; $Y_i$ - output at the position on the top plate, i = 1, 2, 3; $D_i$ - dynamic between the shaker and transducer; $T_{mn}$ - dynamic between transducer m and n, m = 1, 2, 3, n = 1, 2, 3.

## 3. Multi-modes MIMO HOPPF

### 3.1 PPF Controller

Caughey and Goh firstly proposed, applied and verified Positive position feedback (PPF) controller using experimental implement in 1982 [16]. According to observing and testing, the PPF controller is insensitive to uncertain natural damping ratios of the structure, which is different from other control laws, found by the researchers [16]. The measurement of position is positive, then it is fed into the compensator. And at the same time, the position signal from the compensator, which is fed back to the structure later, is also positive [18].

This character makes the PPF controller quite fit for collocated actuator/sensor pairs.

The structure and compensator equations 3.1 and 3.2 are show in the scalar case [16]:

Structure: $\ddot{\varepsilon} + 2\xi\omega\dot{\varepsilon} + \omega^2\varepsilon = g\omega^2\eta$

(3.1)

Compensator: $\ddot{\eta} + 2\xi_f\omega_f\dot{\eta} + \omega_f^2\eta = \omega_f^2\varepsilon$

(3.2)

where g is the scalar gain (positive), $\varepsilon$ is the modal coordinate (structural), $\eta$ is the filter coordinate (electrical), $\omega$ are the structural frequencies, $\omega_f$ are the filter frequencies, $\zeta$ are the structural damping ratios and $2\xi_f$ are the filter damping ratios. This non-dynamic stability criterion is characteristic of the positive position feedback system.

As the second-order transfer function, PPF compensator can be seen in equation (3.3). The transfer function form will be used in this text for deriving the properties of the control system [16].

$$G_{ppf}(s) = \frac{g\omega_f^2}{s^2 + 2\xi_f\omega_f s + \omega_f^2}$$

(3.3)

Another feature of the compensator equation, its second-order low-pass characteristic, led to the terminology PPF filter. Different from a mechanical vibration absorber , the PPF filter behaves much like an electronic vibration absorber for the structure [1]. The PPF controllers only requires one second-order term to repress one vibration mode. So the designer can choose the modes, which are needed to be controlled in a specific bandwidth [60].

The feedback controllers for multivariate resonant systems can be written as [60,1]:

$$G(s) = \sum_{i=1}^{M} \frac{\psi_i \psi_i'}{s^2 + 2\xi_i \omega_i s + \omega_i^2}$$

(3.4)

where $\psi_i$ is an mx1 vector, and M $\rightarrow$ ∞. In practice, we hope that the integer is finite. This is because that only including a very large number of modes, that equation can be sufficiently described the excitation of elastic structure.

In order to deliver stability conditions for this control loop, the series in (3.4) is truncated by keeping the first N modes (N<M). These N modes are in the bandwidth of interest [60,61]. A feed-through term is added to the truncated model to remedy the effect of truncated modes [60,61]. The series in (3.4) are approximated as [60,61]:

$$G^N(s) = \sum_{i=1}^{N} \frac{\psi_i \psi_i'}{s^2 + 2\xi_i \omega_i s + \omega_i^2} + D$$

(3.5)

## 3.2. Multi-modes MIMO HOPPF Controller Parameter Selection

Based the dynamics model between shaker and plate, the results of multi-modes MIMO HOPPF controller optimal parameters are shown in Fig3.1, which are found through GA searching minimum value $H_\infty$ norm of feedback closed-loop system using MATLAB toolbox.

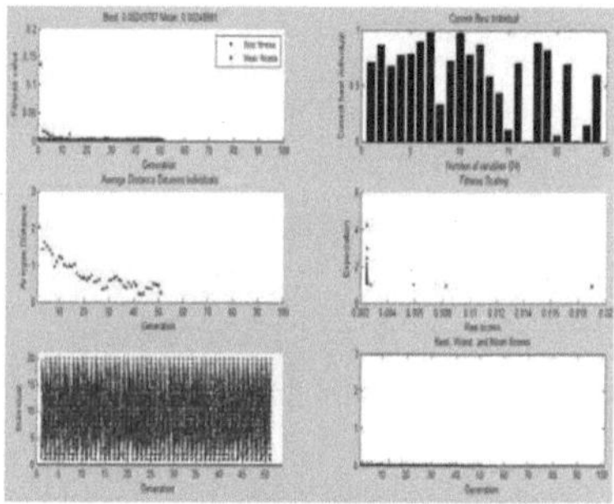

Figure 3.1 optimal parameter results through GA

## 4. Simulation

The multi-modes MIMO HOPPF controller is used to control the first four vibration modes of the plate structure simulation model using mathematical software MATLAB.

Firstly, plate structure simulation model and multi-modes MIMO HOPPF controller are connected as feedback closed-loop system.

Secondly, dynamics model from the shaker to the transducer is connected to feedback closed-loop system.

Thirdly, the parameters of the controller are chosen from the GA searching result.

Fourthly, sweep signal, which contains resonant frequencies of the first four modes of the plate structure simulation model, is applied to the shaker.

Fifthly, Open-loop dynamics, which is between shaker to plate without the controller, is compared to closed-loop dynamics, which is between shaker to plate with the controller.

Time domain and frequency domain cases are designed to test the ability of the multi-modes MIMO HOPPF controller to attenuate first four modes vibration when through observing, the controller centre frequencies and the resonant frequencies of the plate are the same.

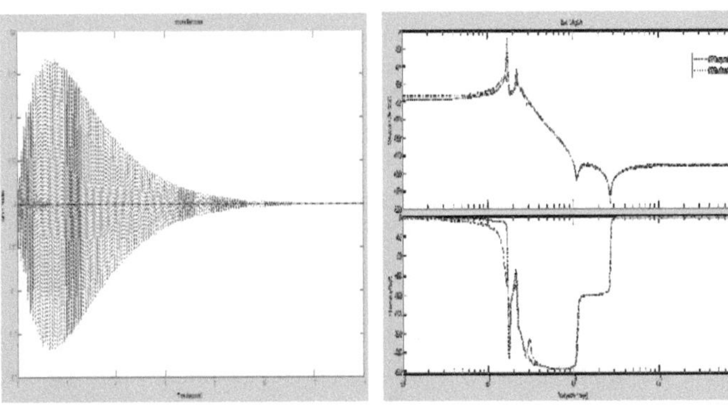

Figure 4.1 open-loop, closed-loop simulation result at transducer 1

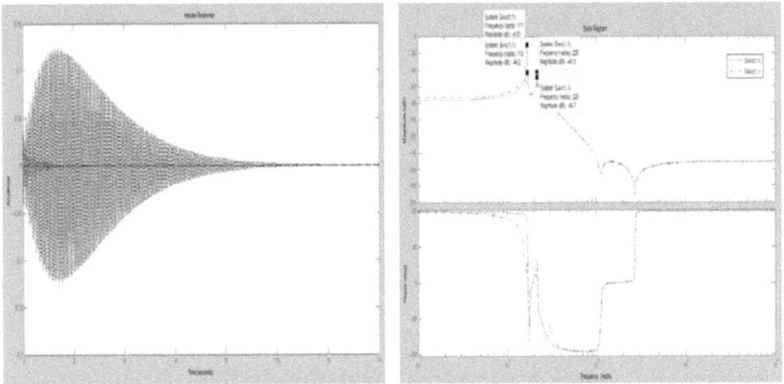

Figure 4.2 open-loop, closed-loop simulation result at transducer 2

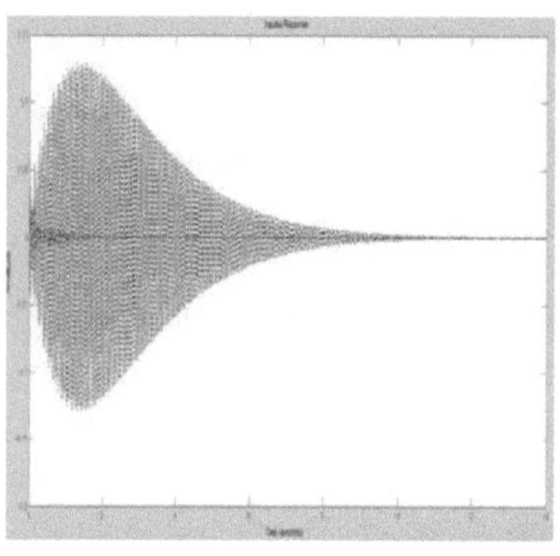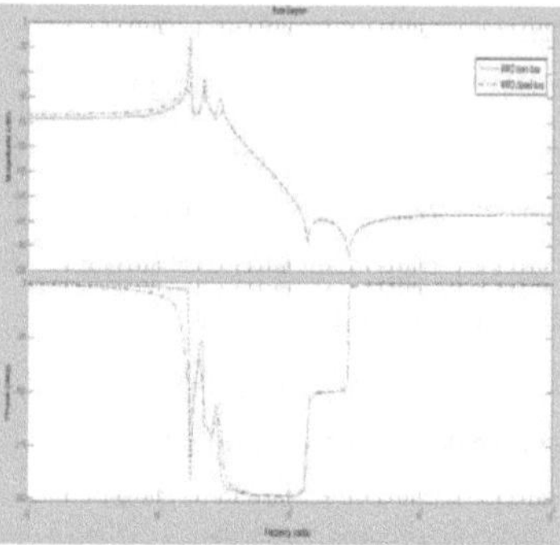

Figure 4.3 open-loop, closed-loop simulation result at transducer 3

## 5. Experiment

The multi-modes MIMO HOPPF controller is tested by author for the first four modes vibration suppression of the plate structure.

A technique called 'self-sensing' was used to measure the back-emf voltage without using any additional sensors. The transducer is performed collocated as it is used for actuation and sensing at the same time [58].

DSpace DS1103 Controller Board was used to interface with the system to be able to acquire the back-emf voltage and to output the reference transducer voltage. 16 16-bit analog-to-digital (AD) converters, 4 12-bit AD converters and 8 14-bit digital-to-analog (DA) converters are provided. Onboard DSP is used for the signal processing and acquisition the output of the various signals via its AD and DA converters when running a simulation externally in MATLAB SIMULINK [58].

Firstly, multi-modes MIMO PPF controller is formed in MATLAB SIMULINK.

Secondly, the sweep signals, which contains resonant frequencies of the first four modes of the plate structure model, are applied to the shaker one by one. The sweep signal is generated from a signal generator, powered and amplified by a 50 W Jay Car amplifier.

Thirdly, a sinusoidal signal, which frequency was chosen from 20 to 50 Hz, was used as the input signal and applied to the shaker, sampling time is 0.1s.

Fourthly, frequency domain cases are designed to test the ability of the multi-modes MIMO HOPPF controller to attenuate vibration of plate structure.

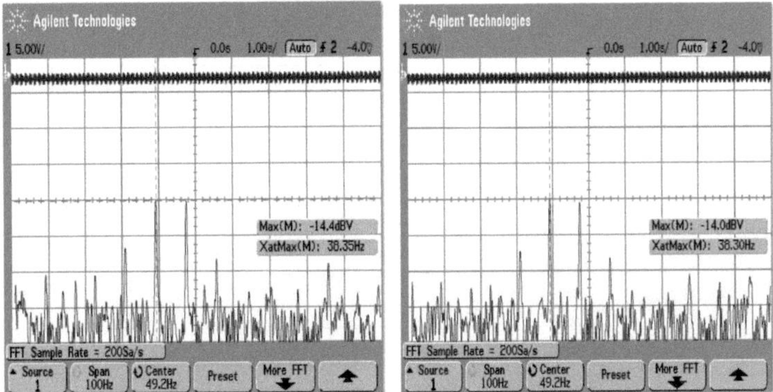

Figure 5.1 at transducer 1 before and after control

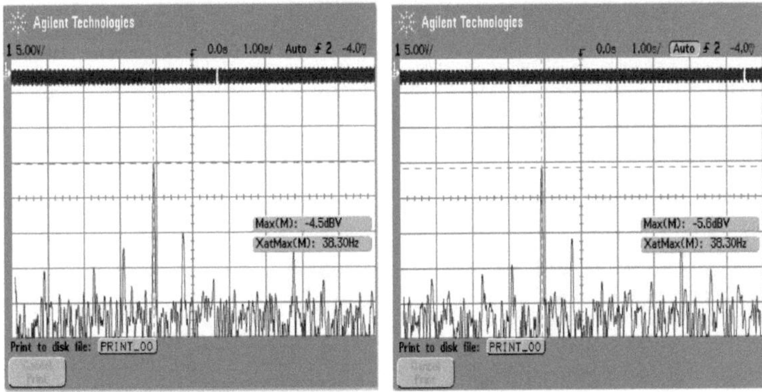

Figure 5.2 at transducer 2 before and after control

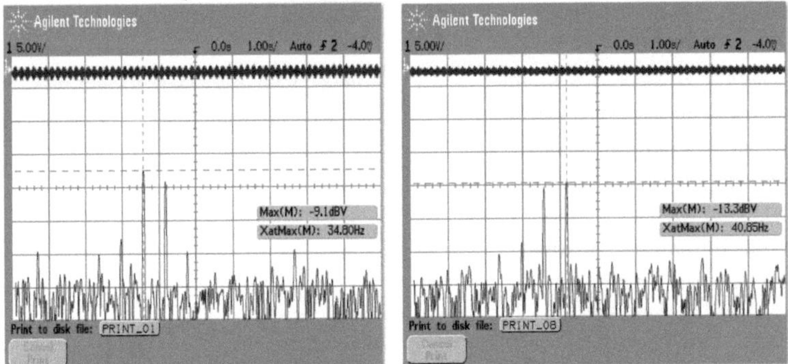

Figure 5.3 at transducer 3 before and after control

| | | | frequency (Hz) | | | |
|---|---|---|---|---|---|---|
| | | | 27.1 | 34.4 | 40.5 | 49.2 |
| transducer 1 | Max Peak (dB) | before | -6.6 | -6.2 | -13.9 | 4.2 |
| | | after | -8.6 | -8.9 | -17.8 | 2.7 |
| | Reduce (dB) | | 2 | 2.7 | 3.9 | 1.5 |
| | Reduce rate (%) | | **19.444%** | **27.64%** | **35.714%** | **14.137%** |
| transducer 2 | Max Peak (dB) | before | -22.6 | -0.8 | -3.1 | -22.5 |
| | | after | -27.6 | -2.8 | -11.5 | -23.5 |
| | Reduce (dB) | | 5 | 2 | 7.6 | 1 |
| | Reduce rate (%) | | 34.01% | 19.27% | 59.26% | 18% |
| transducer 3 | Max Peak (dB) | before | -20.3 | -2.3 | -4.3 | -20.4 |
| | | after | -21.3 | -5.7 | -11.2 | -22.4 |
| | Reduce (dB) | | 1 | 3.4 | 6.9 | 2 |
| | Reduce rate (%) | | **19.33%** | **33.33%** | **50.543%** | **29.091%** |

Table 5.1    multi-modes MIMO HOPPF vibration control experimental results

## 6. Summary

According to simulation and experiment result, the provided methodology multi-modes MIMO HOPPF controller can be successfully used for the active vibration suppression of plate structure. The author achieved good vibration suppression results at first four modes of the plate. Compared to other methods, there is no noise happened at the beginning of turning on the controller and there is no spillover.


## Acknowledgement

S.S. Groothuis and R.J. Roesthuis contributed to design the experiment plant and ANASYS simulation model in 2010. Siyang Yu contributed to modify the ANSYS simulation model in 2012.


## Conflict of interest statement

The Author declares that there is no conflict of interest.

**References**

[1] Tjahyadi, H. (2007) Adaptive multi mode vibration control of dynamically loaded flexible structures (Doctoral dissertation, Flinders University)

[2] Li, S. (2011). Active modal control simulation of vibro-acoustic response of a fluid-loaded plate. Journal of Sound and Vibration, 330(23), 5545-5557

[3] Kim, S. M., & Oh, J. E. (2013). A modal filter approach to non-collocated vibration control of structures. Journal of Sound and Vibration

[4] Meirovitch, L., Van Landingham, H. F., & Öz, H. (1977). Control of spinning flexible spacecraft by modal synthesis. Acta Astronautica, 4(9), 985-1010

[5] Meirovitch, L., Baruh, H., & OZ, H. (1983). A comparison of control techniques for large flexible systems. Journal of Guidance, Control, and Dynamics, 6(4), 302-310

[6] Iwamoto, H., Tanaka, N., & Hill, S. G. (2012). Feedback control of wave propagation in a rectangular panel, part 2: Experimental realization using clustered velocity and displacement feedback. Mechanical Systems and Signal Processing

[7] Moheimani, S. O. R., Fleming, A. J., & Behrens, S. (2001). Highly resonant controller for multimode piezoelectric shunt damping. Electronics Letters, 37(25), 1505-1506

[8] Pota, H. R., Moheimani, S. R., & Smith, M. (2002). Resonant controllers for smart structures. Smart Materials and Structures, 11(1), 1

[9] Aphale, S. S., Fleming, A. J., & Moheimani, S. R. (2007). Integral resonant control of collocated smart structures. Smart Materials and Structures, 16(2), 439

[10] Pereira, E., Moheimani, S. O. R., & Aphale, S. S. (2008). Analog implementation of an integral resonant control scheme. Smart Materials and Structures, 17(6), 067001.

[11] Mahmood, I. A., Moheimani, S. R., & Bhikkaji, B. (2008). Precise tip positioning of a flexible manipulator using resonant control. Mechatronics, IEEE/ASME Transactions on, 13(2), 180-186

[12] Pereira, E., Aphale, S. S., Feliu, V., & Moheimani, S. R. (2011). Integral resonant control for vibration damping and precise tip-positioning of a single-link flexible manipulator. Mechatronics, IEEE/ASME Transactions on, 16(2), 232-240

[13] Halim, D., & Moheimani, S. R. (2001). Spatial resonant control of

flexible structures-application to a piezoelectric laminate beam. Control Systems Technology, IEEE Transactions on, 9(1), 37-53

[14]Moheimani, S. R., Vautier, B. J., & Bhikkaji, B. (2006). Experimental implementation of extended multivariable PPF control on an active structure. Control Systems Technology, IEEE Transactions on, 14(3), 443-455

[15]Goh, C. J. (1983). Analysis and control of quasi distributed parameter systems (Doctoral dissertation, California Institute of Technology)

[16]Goh, C. J., & Caughey, T. K. (1985). On the stability problem caused by finite actuator dynamics in the collocated control of large space structures. International Journal of Control, 41(3), 787-802

[17]Fanson J L, An experimental investigation of vibration suppression in large space structures using positive position feedback, PhD Thesis California Institute of Technology,November,1986

[18]J. L. Fanson, and T. K. Caughey, "Positive Position Feedback Control for Large Space Structures," AIAA Journal, vol. 28, no. 4, pp.717-724, 1990

[19]Song G, Schmidt S P and Agrawal B N, Experimental robustness study of positive position feedback control for active vibration suppression, J.

Guidance, VOL. 25, NO. 1: ENGINEERING NOTES, 2001

[20]J. L. Fanson, and T. K. Caughey, "Positive Position Feedback Control for Large Space Structures," AIAA Journal, vol. 28, no. 4, pp.717-724, 1990

[21]Song G, Schmidt S P and Agrawal B N, Experimental robustness study of positive position feedback control for active vibration suppression, J. Guidance, VOL. 25, NO. 1: ENGINEERING NOTES, 2001

[22]Baz, A., Poh, S., & Fedor, J. (1989). Independent modal space control with positive position feedback. Dynamics and control of large structures, 553-567

[23]Poh, S., & Baz, A. (1990). Active control of a flexible structure using a modal positive position feedback controller. Journal of Intelligent Material Systems and Structures, 1(3), 273-288.

[24]Goh, C. J., & Lee, T. H. (1991). Adaptive modal parameters identification for collocated position feedback vibration control. International Journal of Control, 53(3), 597-617.

[25]Caughey, T. K. (1995). Dynamic response of structures constructed from smart materials. Smart Materials and Structures, 4(1A), A101

[26]Baz, A., & Poh, S. (1996). Optimal vibration control with modal positive position feedback. Optimal

control applications and methods, 17(2), 141-149

[27] Baz, A., & HONG, J. T. (1997). Adaptive control of flexible structures using modal positive position feedback. International journal of adaptive control and signal processing, 11(3), 231-253

[28] Wang, L. (2003, October). Positive position feedback based vibration attenuation for a flexible aerospace structure using multiple piezoelectric actuators. In Digital Avionics Systems Conference, 2003. DASC'03. The 22nd (Vol. 2, pp. 7-C). IEEE

[29] Shan, J., Liu, H. T., & Sun, D. (2005). Slewing and vibration control of a single-link flexible manipulator by positive position feedback (PPF). Mechatronics, 15(4), 487-503

[30] Hong, C., Shin, C., & Jeong, W. (2010). Active vibration control of clamped beams using PPF controllers with piezoceramic actuators. Proc. 20th Int. Congr. on Acoustics (Sydney, Australia, ICA 2010 23–27 August 2010)

[31] Ahmed, B., & Pota, H. R. (2011). Dynamic compensation for control of a rotary wing UAV using positive position feedback. Journal of Intelligent & Robotic Systems, 61(1-4), 43-56

[32] Chuang, K. C., Ma, C. C., & Wu, R. H. (2012). Active suppression of a beam under a moving mass using a pointwise fiber bragg grating displacement sensing system. Ultrasonics, Ferroelectrics and Frequency Control, IEEE Transactions on, 59(10), 2137-2148

[33] Shin, C., Hong, C., & Jeong, W. B. (2012). Active vibration control of clamped beams using positive position feedback controllers with moment pair. Journal of mechanical science and technology, 26(3), 731-740

[34] Bang, H., & Agrawal, B. N. (1994). A generalized second order compensator design for vibration control of flexible structures. In AIAA/ASME/ASCE/AHS/ASC Structures, Structural Dynamics, and Materials Conference, 35th, Hilton Head, SC (pp. 2438-2448)

[35] Mahmoodi, S. N., Aagaah, M. R., & Ahmadian, M. (2009, June). Active vibration control of aerospace structures using a modified positive position feedback method. In American Control Conference, 2009. ACC'09. (pp. 4115-4120). IEEE

[36] Hu, Q., Xie, L., & Gao, H. (2006, December). A combined positive position feedback and variable structure approach for flexible spacecraft under input nonlinearity. In Control, Automation, Robotics and Vision, 2006. ICARCV'06. 9th International Conference on (pp. 1-6). IEEE

[37] Hu, Q., Xie, L., & Gao, H. (2007). Adaptive variable structure and active vibration reduction for flexible spacecraft under input nonlinearity. Journal of Vibration and Control, 13(11), 1573-1602

[38] Kim, S. M., Wang, S., & Brennan, M. J. (2011). Comparison of negative and positive position feedback control of a flexible structure. Smart Materials and Structures, 20(1), 015011

[39] Bhikkaji, B., Ratnam, M., Fleming, A. J., & Moheimani, S. R. (2007). High-performance control of piezoelectric tube scanners. Control Systems Technology, IEEE Transactions on, 15(5), 853-866

[40] Chen, L., He, F., & Sammut, K. (2002). Nonlinear modal positive position feedback for vibration control in distributed parameter systems. In Annual Conference

[41] Qiu, Z. C., Zhang, X. M., Wu, H. X., & Zhang, H. H. (2007). Optimal placement and active vibration control for piezoelectric smart flexible cantilever plate. Journal of Sound and Vibration, 301(3), 521-543

[42] Song, G., Qiao, P. Z., Binienda, W. K., & Zou, G. P. (2002). Active vibration damping of composite beam using smart sensors and actuators. Journal of Aerospace Engineering, 15(3), 97-103

[43] Song G, Schmidt S P and Agrawal B N, Experimental robustness study of positive position feedback control for active vibration suppression, J. Guidance, VOL. 25, NO. 1: ENGINEERING NOTES, 2001

[44] Long, Z., & Guangren, D. (2012, May). A robust vibration suppression controller design for space flexible structures. In Control and Decision Conference (CCDC), 2012 24th Chinese (pp. 4020-4024). IEEE

[45] Han, J. H., & Lee, I. (1999). Optimal placement of piezoelectric sensors and actuators for vibration control of a composite plate using genetic algorithms. Smart Materials and Structures, 8(2), 257

[46] Qiu, Z. C., Zhang, X. M., Wu, H. X., & Zhang, H. H. (2007). Optimal placement and active vibration control for piezoelectric smart flexible cantilever plate. Journal of Sound and Vibration, 301(3), 521-543

[47] Moheimani, S. O. R., Vautier, B. J. G., & Bhilkkaji, B. (2005, December). Multivariable PPF control of an active structure. In Decision and Control, 2005 and 2005 European Control Conference. CDC-ECC'05. 44th IEEE Conference on (pp. 6824-6829). IEEE

[48] Orszulik, R. R., & Shan, J. (2012). Active vibration control using genetic algorithm-based system identification and positive position

feedback. Smart Materials and Structures, 21(5), 055002

[49] Kwak, M. K., & Han, S. B. (1998, July). Application of genetic algorithms to the determination of multiple positive-position feedback controller gains for smart structures. In 5th Annual International Symposium on Smart Structures and Materials (pp. 637-648). International Society for Optics and Photonics

[50] Kwak, M. K., & Shin, T. S. (1999, June). Real-time automatic tuning of vibration controllers for smart structures by genetic algorithm. In Proceedings of SPIE- The International Society for Optical Engineering (Vol. 3667, pp. 679-690)

[51] Kwak, M. K., & Heo, S. (2000, June). Real-time multiple-parameter tuning of PPF controllers for smart structures by genetic algorithms. In SPIE's 7th Annual International Symposium on Smart Structures and Materials (pp. 279-290). International Society for Optics and Photonics

[52] Kwak, M. K., & Heo, S. (2007). Active vibration control of smart grid structure by multiinput and multioutput positive position feedback controller. Journal of Sound and Vibration, 304(1), 230-245

[53] Han, J. H., & Lee, I. (1999). Optimal placement of piezoelectric sensors and actuators for vibration control of a composite plate using genetic algorithms. Smart Materials and Structures, 8(2), 257

[54] Kwak, M. K., Heo, S., & Jeong, M. (2009). Dynamic modelling and active vibration controller design for a cylindrical shell equipped with piezoelectric sensors and actuators. Journal of Sound and Vibration, 321(3), 510-524

[55] Ojeda, X., Mininger, X., Gabsi, M., & Lécrivain, M. (2008). Noise cancellation of 6/4 switched reluctance machine by piezoelectric actuators: Optimal design and placement using genetic algorithm

[56] Qiu, Z. C., Wu, H. X., & Ye, C. D. (2009). Acceleration sensors based modal identification and active vibration control of flexible smart cantilever plate. Aerospace Science and Technology, 13(6), 277-290

[57] Qiu, Z. C., Zhang, X. M., Wu, H. X., & Zhang, H. H. (2007). Optimal placement and active vibration control for piezoelectric smart flexible cantilever plate. Journal of Sound and Vibration, 301(3), 521-543

[58] S.S. Groothuis and R.J. Roesthuis, February 2010. Vibration Cancellation Using Synthetic Shunt Impendances. Flinders University, Adelaide, Australia

[59] Moheimani, S. R., Halim, D., & Fleming, A. J. (2003). Spatial

Control of Vibrations: Theory and Experiments. World scientific

[60] Halim, D., & Moheimani, S. O. (2004). Reducing the effect of truncation error in spatial and pointwise models of resonant systems with damping. Mechanical systems and signal processing, 18(2), 291-315

[61] S. O. R. Moheimani. Experimental verification of the corrected transfer function of a piezoelectric laminate beam. IEEE Transactions on Control Systems Technology, 8(4):660-666, July 2000

[62] Omidi, E., & Mahmoodi, S. N. (2014). Vibration control of collocated smart structures using H∞ modified positive position and velocity feedback. Journal of Vibration and Control, 1077546314548471.

[63] Ferrari, G., & Amabili, M. (2015). Active vibration control of a sandwich plate by non-collocated positive position feedback. Journal of Sound and Vibration, 342, 44-56.

[64] Omidi, E., Mahmoodi, S. N., & Shepard, W. S. (2015). Vibration reduction in aerospace structures via an optimized modified positive velocity feedback control. Aerospace Science and Technology, 45, 408-415.

# YOUR KNOWLEDGE HAS VALUE

- We will publish your bachelor's and
  master's thesis, essays and papers

- Your own eBook and book -
  sold worldwide in all relevant shops

- Earn money with each sale

Upload your text at www.GRIN.com
and publish for free